What am I?
Fruits

Written and illustrated by John Benzee

What am I? Fruits

Copyright © 2019 by John Benzee
All illustrations and text by John Benzee

All rights reserved.
No part of this book may be used or reproduced in any manner whatsoever without permission, except in the case of brief quotations in critical articles or reviews.

Published by Split Seed Press; Clarence, NY
ISBN: 978-0-9997379-3-4 (hardcover)
ISBN: 978-0-9997379-4-1 (paperback)
Visit johnbenzee.com for more information

First edition
Typeset in Roboto

Publisher's Cataloging-In-Publication Data:
Names: Benzee, John, 1995— author, illustrator.
Title: What am I? Fruits / by John Benzee.
Description: Clarence : Split Seed Press, 2019. | Series: What am I? series. | Summary: A series of riddles that give descriptions of various fruits and then reveal their names.
Identifiers: ISBN 9780999737934 (hardcover) | ISBN 9780999737941 (paperback)
Subjects: LCSH: Fruit—Varieties—Juvenile literature. | Riddles, Juvenile.
Classification: SB357.2.B46 2019 | DDC 634--dc23

10 9 8 7 6 5 4 3 2 1

Note: This book follows the USDA, not botanical, fruit and vegetable classification.

I am a variety of fruits
Grown in an orchard, field, or garden.

I have some riddles for you,
Curious reader,
To figure out
What types of fruits I am.

An Apple? A lemon? Grapes?

Turn the page to start guessing.

I'm a blue colored marble,
Squishy to eat,
With a flavor that's lightly sweet.

I grow on a bush
In tight clusters,
My skin has a purple luster.

My berries ripen
In the summer sun,
Ready for some picking fun.

What am I?

I'm a June berry
Grown close to the ground.
My shape isn't perfectly round.

From white flowers
I grow a bright red fruit.
I can send out side shoots.

My yellow seeds
Cling to my outside,
With a juicy, sweet inside.

What am I?

I grow on a
Long, woody vine,
With tendrils that like to twine.

In big clusters
My oval fruit grow,
Sometimes seeds I forgo.

I'm deep purple
Or bright green.
I am perfect to glean.

What am I?

I'm grown on a tree
And shaped like a ball,
Ready to eat by fall.

My waxy skin
Is yellow, green, or red,
With a strong stem overhead.

My white, crispy flesh
Surrounds a seeded core.
I can be eaten, juiced, or much more.

What am I?

I'm a sweet, but gritty fruit
Grown on a tree,
About the size of a cup of tea.

Off the tree,
My green fruit turns tender,
Then taste my moist splendor.

Colored dark yellow,
Reddish-brown or green,
My tear-shape can be seen.

What am I?

In summer I grow
On a long, green vine.
My number of fruit is less than nine.

I'm often striped
And bigger than a bowling ball.
I expand quickly when I'm small.

My red, watery flesh,
With seeds crunchy and black,
Is a perfect sliced snack.

What am I?

On a tree I grow,
Red-purple in color,
Bigger than a sand dollar.

My hard outer skin,
Cut into sections of four,
Reveals my inner, seeded core.

I'm eaten for my seeds
Or drunk as healthy juice.
My inedible peel isn't much use.

What am I?

My red, bumpy berry
Is hollow inside,
About a half-inch wide.

Watch out for thorns,
For I grow on long canes.
My sweetness is worth the labor pains.

What am I?

I'm a berry cousin
Who is dark purple or black,
An attached stem I don't lack.

On bushy brambles
My quarter-sized berries grow,
With seeded clusters found in rows.

What am I?

Shaped like a smile,
I'm bright yellow.
I may soon brown and mellow.

On a tall, leafy plant
I ripen from green,
Growing in clusters I can be seen.

To eat me up
Remove my peel,
For a soft, tropical meal.

What am I?

I am a stone fruit
With a hard oval pit,
Fuzzy skin is my outfit.

My red-yellow flesh
Is filled with sweet juice,
Hunger I can induce.

I grow on a tree,
With flower petals colored pink,
Gobble me up in a wink.

What am I?

I grow on a flowery tree
And dangle from a stem.
I am nature's sweet gem.

A small hard pit
Can be found inside,
While my dark red skin is my pride.

My quarter-sized fruit
Is perfect to eat,
Especially atop a sundae treat.

What am I?

I'm a citrus,
Orange and bright,
A lover of daylight.

Slice me up
Or peel back my rind,
Segments you will find.

I'm found on a tree
In the tropical heat.
My juice is tasty and sweet.

What am I?

I grow on a woody plant
In the hot sun.
My many flowers turn to one.

My yellow, inner flesh
Is protected by brown scales outside.
Wait six months 'til I can be tried.

With a green crown,
Leaves waxy and tough,
Flowering once is enough.

What am I?

On a tree I grow,
Round and bright yellow,
My sour taste makes you mellow.

Commonly used
Are my juice, rind, or peel
Or added as flavoring to some meals.

My juice is drunk,
A sugar and water mix,
For a refreshing summer fix.

What am I?

I'm a juicy fruit,
Shaped like a globe,
My skin has a waxy robe.

Commonly dark red or purple,
Centered around an oval pit,
My tart flavor completes my outfit.

Eat me fresh,
Or dried as a prune,
Pick me before the harvest moon.

What am I?

About the fruits

 **BLUEBERRIES** grow on bushes. They start off green and then ripen to blue.

 **STRAWBERRIES** have seeds on the outside. They send out runners to make new plants.

 **GRAPES** commonly grow in vineyards on woody vines. They come in seed and seedless types.

 **APPLES** are widely grown and have a core with seeds. There are many varieties.

 **PEARS** soften when ripe. Their flesh contains stone cells, called grit.

 **WATERMELONS** grow on trailing vines. The fruit slowly expands as it absorbs water.

 **POMEGRANATES** have a hard husk that conceals its soft, red arils, which contain a seed and juice.

 **RASPBERRIES** are a cluster of seeds (drupelets). **BLACKBERRIES** are similar, but the stem remains on the berry when picked.

 **BANANAS** grow in clusters on tall plants. When overripe, their yellow peel turns brown.

 **PEACHES** have fuzzy, red-orange skin. Inside, they have a large pit (a seed).

 **CHERRIES**, sweet or sour, grow in clusters. Inside, they have a small pit.

 **ORANGES** have rinds that cover the segmented, squishy membrane. There are many varieties/ hybrids.

 **PINEAPPLES** form when its berries join together to produce one fruit. It usually only flowers once.

 **LEMONS** have sour juice, which can be used as flavoring or made into lemonade.

 **PLUMS** are a stone fruit (hard pit). Their skin is covered in a natural, whitish wax for protection.

www.ingramcontent.com/pod-product-compliance
Lightning Source LLC
Chambersburg PA
CBHW051555010526
44118CB00022B/2712